AIChE Equipment Testing Procedure

CENTRIFUGAL PUMPS

(Newtonian Liquids)

A Guide to Performance Evaluation

Second Edition

Prepared by the
Equipment Testing Procedures Committee

345 East 47 Street, New York, NY 10017

TABLE OF CONTENTS

Printed in the United States of America by
TWIN PRODUCTION & DESIGN

ISBN 0-8169-0320-4

AIChE EQUIPMENT TESTING PROCEDURE FOR CENTRIFUGAL PUMPS

100.0 PURPOSE AND SCOPE

101.0 *Purpose*

101.1 This testing procedure suggests methods of conducting and interpreting field tests on centrifugal pumps with actual pumped fluids. Measuring methods practical for field test conditions are presented to guide the user. These tests may be desired for the following purposes:

101.1.1 Checking a new pump for operability, to determine if a pump functions both mechanically and hydraulically to do the job for which it was intended.

101.1.2 Testing a pump for overall system performance, to determine if the pump will deliver all the fluid which it was designed to deliver under various load conditions as demanded by the process.

101.1.3 Testing to establish operating characteristics, such as total head, capacity, and power requirements. Net Positive Suction Head Required may be determined as part of this procedure when test code accuracy is not required. Such measurements can also help resolve operating problems.

101.1.4 Testing a pump as part of a regularly scheduled maintenance program. An indication of the condition of the pump internals can be obtained.

101.1.5 Periodic testing to establish a performance history which can be used to monitor a change in the pumping conditions, and energy consumption. Performance test can be used to determine the mechanical condition of the pump or if a change in the process has occurred.

102.0 *Scope*

102.1 This procedure recommends methods for conducting and reporting field tests on centrifugal pumps handling Newtonian fluids.

102.2 It should be noted that the preferred method of establishing pump performance is under controlled laboratory conditions. Testing can generally be accomplished by the pump manufacturer at the factory and may be witnessed by the purchaser. It should also be noted that this field test procedure should not be used when test code accuracy is required.

102.3 Safety and environmental concerns are listed in Sections 301.1 and 301.2.

200.0 DEFINITIONS AND DESCRIPTION OF TERMS

201.0 This section contains the definition of all terms used in this testing procedure. Nomenclature and typical units are also listed in Section 801.0.

202.0 *Centrifugal Pumps:* A pump in which pressure is developed principally by the action of centrifugal force.

203.0 *Newtonian Fluids:* A liquid which resists shear in direct proportion to the rate of shear.

204.0 *Datum:* The reference line from which all elevations are measured.

205.0 *Specific Gravity, (s):* The dimensionless ratio of the specific weight of the liquid to the specific weight of water under reference conditions of temperature and pressure. It is a function of temperature, operating pressure, and elevation.

206.0 *Vapor Pressure, (h_{vpa}):* The absolute pressure at which the liquid will change to vapor at a given temperature.

207.0 *Viscosity, (μ):* The measure of a fluid's tendency to resist an internal shearing force.

208.0 *Pressure Head, (h_p):* The pressure energy per unit weight of liquid at a given location corrected to datum elevation.

209.0 *Total Discharge Head, (h_d):* The reading of a pressure gauge at the discharge of the pump, converted to feet of liquid and referred to datum, plus the velocity head at the point of gauge attachment.

210.0 *Total Suction Head, (h_s):* The reading of a pressure gauge at the suction of the pump, converted to feet of liquid and referred to datum, plus the velocity head at the point of gauge attachment.

210.1 *Total Suction Lift, (h_{gs}):* Suction lift exists where the total suction head is below atmospheric pressure. Total suction lift, as determined during testing, is the reading of a liquid manometer at the suction nozzle of the pump, converted to feet of liquid and referred to datum, minus the velocity head at the point of gauge attachment.

211.0 *Pump Total Head, (H):* The measure of the work increase, per pound of the liquid, imparted to the liquid by the pump. Therefore, pump total head is the algebraic difference between the total discharge head and total suction head.

212.0 *Static Discharge Head, (h_{sd}):* The hydraulic pressure exerted by the liquid against the pump discharge as indicated by a pressure gauge, manometer or similar devices, referred to datum and converted to feet of the liquid being pumped.

213.0 *Static Suction Head (h_{ss}):* The hydraulic pressure exerted by the liquid against the pump suction as indicated by a pressure gauge, manometer or similar devices, referred to datum and converted to feet of the liquid being pumped.

214.0 *Suction Surface Pressure, (h_{ps}):* The pressure exerted on the surface of the liquid at the suction sources of the pump.

215.0 *Friction Head Loss, (h_f):* The resistance to flow in a pumping system as indicated by loss of head.

216.0 *Gravitational Acceleration, (g):* The gravitational constant is 32.2 ft/sec^2 at sea level.

217.0 *Velocity Head, (h_v):* The energy component in a pumping system that represents the kinetic or "velocity" energy in a moving liquid at the point being considered in the system. It is equivalent to the vertical distance the mass of liquid would have to fall (in a perfect vacuum) to acquire the velocity V and is expressed as:

$$h_v = \frac{V^2}{2g}$$

218.0 *Pump Efficiency, (Eff):* The output (hydraulic) horsepower supplied by the pump divided by the input (shaft) horsepower expressed as a percentage.

219.0 *Net Positive Suction Head Required,* ($NPSH_R$): Net Positive Suction Head Required ($NPSH_R$) is a crucial performance limitation of any centrifugal pump. It must be considered with equal importance to head, capacity, and working pressure when selecting a pump for a given service. The $NPSH_R$ of a given pump is a function of impeller design and is determined by testing. The $NPSH_R$ curve furnished by the manufacturer is the minimum required by the pump to maintain stable operation and is based on a datum elevation, usually the centerline of the impeller. A pump, in order to operate free of cavitation, must be supplied with $NPSH_A$ at least equal to, preferably in excess of, the $NPSH_R$ value.

219.1 *Net Positive Suction Head Available,* ($NPSH_A$): Net Positive Suction Head Available ($NPSH_A$) is a function of the suction system feeding the pump. A detailed calculation of $NPSH_A$ should have been done prior to selecting the pump and the test results should confirm there is a surplus over the $NPSH_R$ of the pump selected.

219.2 To calculate, $NPSH_A$ equals the suction surface pressure at the liquid surface, minus the vapor pressure, plus the suction static head (all in feet of liquid) minus the friction loss through the suction system. If suction pressure is measured at the pump suction flange, the friction loss (h_{fs}) value is zero.

$$NPSH_A = h_{psa} + h_{ss} - h_{fs} - h_{vpa},$$

h_{psa} = suction surface pressure, absolute

h_{ss} = static suction head (plus or minus depending on actual conditions)

h_{fs} = suction friction head

h_{vpa} = vapor pressure of liquid, absolute, at temperature.

219.3 Several common $NPSH_A$ situations are shown in Figures 219.3.1, 219.3.2, and 219.3.3.

219.3.1 Pressurized tank on suction

$$NPSH_A = 2.31 \frac{h_{psa} - h_{vpa}}{s} + h_{ss} - h_{fs}$$

s = specific gravity of liquid at pumping.

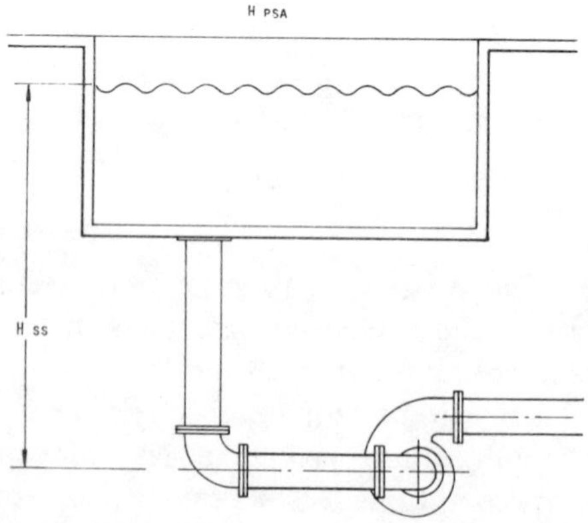

Figure 219.3.1 Pressurized tank on suction.

219.3.2 Pulling a suction lift

$$NPSH_A = 2.31 \frac{h_{psa} - h_{vpa}}{s} - h_{ss} - h_{fs}$$

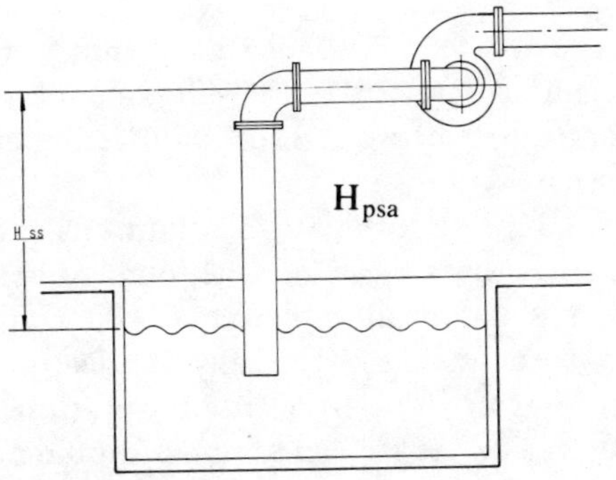

Figure 219.3.2 Pulling a suction lift.

Note: h_{ps} = atmospheric pressure on an open tank

219.3.3 Liquid at boiling point

$$NPSH_A = 2.31\frac{h_{psa} - h_{vpa}}{s} + h_{ss} - h_{fs}$$

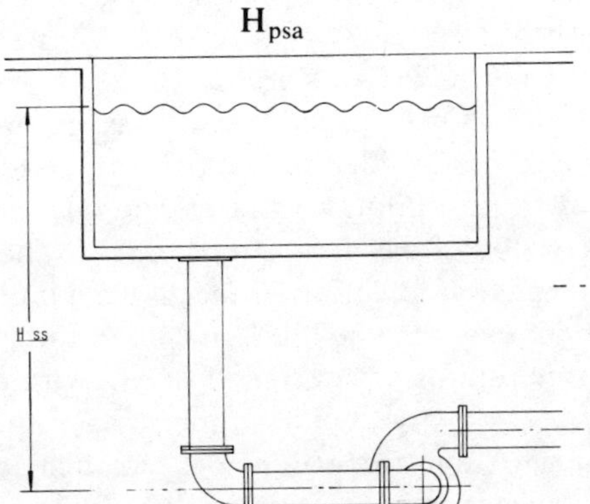

Figure 219.3.3 Liquid at boiling point.

Note: But for boiling liquids $h_{psa} = h_{vpa}$, *therefore:*

$$NPSH_A = h_{fs} + h_{ss}$$

300.0 TEST PLANNING

301.0 *Conditions*

301.1 Safety: Any equipment testing must conform to the latest requirements of all applicable safety standards. These include but are not limited to plant, industry, Local, State, and Federal regulations. It is recommended that all testing be conducted under the supervision of personnel fully experienced in plant and equipment operating practices.

301.2 *Environmental:* The test procedures must conform to the latest requirements of all applicable environmental standards which include plant, industry, Local, State, and Federal regulations. Environmental conditions that apply to the equipment in normal operation should also apply during testing.

301.3 *Liability:* AIChE and members of the various committees involved make no representation, warranties or guarantees, expressed or implied, as to the application or fitness of the testing procedures suggested herein for any specific purpose or use. Company affiliations are shown for information only and do not imply procedure approval by the companies listed. The user ultimately must make his own judgment as to the testing procedures he wishes to utilize for specific applications.

302.0 *Resources*

302.1 *Time:* The test should be scheduled at a time when suitable operating conditions can be obtained and the presence of key personnel to carry out the test can be assured. Time should be allowed to determine objectives and provide proper instrumentation before an actual test is performed.

302.2 *Equipment:* After determining which measurements are necessary for achieving the test objectives, equipment or instrumentation must be selected and installed to obtain the necessary measurements. The required precision or accuracy of the measurements is an important factor in determining which instruments should be used. Also, proper consideration should be given to calibrating the instruments and installing them according to recognized procedures or practices.

303.0 *Process Considerations for Performing Tests*

If the test is to be performed under actual process conditions, as in an operating production unit, attention should be given to possible process upsets. A test may require the flow be "dead headed" (no flow) or that maximum flow rates be obtained. Also, suction and discharge pressures may be varied during a test. Determination must be made if the process can tolerate large changes in conditions, such as flow and pressure, without having the process get out of control. Operating the pump at various flows could also change the temperature of the liquid being pumped. Having to operate the pump at conditions other than at design can cause mechanical problems that could result in a sudden shutdown of the pump, i.e., motor overload. Knowing the capability of the pump and the process system is essential in setting up the test to avoid unwanted upsets.

304.0 *Data Requirements*

The data required to evaluate the perfomance of a centrifugal pump depend on the purpose of the test. The following data may be required.

304.1 *Field Test Data*

304.1.1 Suction pressure at or near the suction flange of the pump.

304.1.2 Discharge pressure at or near the discharge flange.

304.1.3 Power Consumption: Electric motor amp draw, electric motor kilowatt usage, steam rate of turbine, or fuel usage of engine.

304.1.4 Capacity (volume rate of flow) at the pump.

304.1.5 Speed of the pump shaft in revolutions per minute.

304.1.6 Liquid temperature at the suction and discharge.

304.1.7 Barometric pressure.

304.2 *Installation Data:* The following data concerning the installation should be known:

304.2.1 Piping details: items such as valves, orifices reducers, suction strainers may affect performance.

304.2.2 Access to the pump.

304.2.3 Proper location of instrumentation. Difference in elevation between the gauges and the centerline of the impeller must be considered.

304.3 *System Data or Equipment Information*

304.3.1 Driver calibration data.

304.3.2 Pressure gauge calibration.

304.3.3 Suction and discharge pipe inside diameters.

304.3.4 Liquid Characteristics: specific gravity, viscosity and temperature effects. Specific gravity, vapor pressure and viscosity measurement are not within the scope of this procedure. These characteristics should be obtained from analyses of process samples or technical reference books.

304.3.5 Pump Information: rated capacity, rated head, rated speed, impeller drawing number, impeller diameter, materials of construction, mechanical seal or packing details. NPSH requirements, efficiency, and performance curve.

305.0 *Pump Considerations*

Certain conditions should be observed and recorded prior to the initiation of a pump performance test.

305.1 The pump should be identified, preferably by serial number. All available information concerning the pump's performance should be obtained. This information will be used later in the test.

305.2 The pump history should be known. A newly installed pump would be tested in a different manner than an older pump. If the pump has had problems, it should be noted prior to the test so that considerations can be given to the effect of the problems on the test results.

305.3 *Pump Installation*

305.3.1 Alignment between pump and driver should be checked.

305.3.2 The installation should be inspected for proper support of the pump. If the support is not adequate, it can cause misalignment between the pump and driver. Misalignment can cause vibration, bearing and seal failure, and coupling wear.

305.3.3 The foundation conditions should be noted at this time. Does the pump appear to be on a substantial concrete foundation? Is it mounted on a substantially fabricated frame? Are there externally excited vibrations in the foundation area? These factors should be noted at this time for later evaluation.

305.3.4 The pump piping system should be observed. Conditions such as long lengths of unsupported pipe, thermal expansion, etc., may impose excessive strain on suction and/or discharge nozzles. Excessive strain could cause misalignment, internal rubbing and affect the performance of the pump.

305.3.5 If a suction strainer is used, the cleanliness and the location may affect pump performance.

305.4 *Hydraulic Conditions*

305.4.1 Operational conditions of the pump should be considered. It is to be noted that there may be flow limitations. Centrifugal pumps may have low flow limitations due to heat rise, internal recirculation or mechanical limitations. There may also be high flow operational limits. Some pumps are designed to run with an open discharge valve while others may require discharge valve throttling to keep them on their performance curve. These possibilities should be checked with the manufacturer.

305.4.2 The suction conditions of the pump should be observed. The suction piping system should be considered to insure proper inlet flow conditions. Observations of sump level should be conducted during the pump test to assure smooth inlet flow. Observation of entrained gas or vortexing should be noted.

305.4.3 In taking performance data, there may be fluctuations in the readings due to pump instability, system instability, or mechanically generated vibrations. These should be identified and corrected before proceeding with the test.

305.4.4 It may be necessary to take data readings simultaneously, such as inlet and outlet temperatures. This may require special provisions if instrumentation is separated from the pump.

305.5 *Conditions That May Affect Performance*

305.5.1 It should be noted that gas in the fluid will affect the performance of a centrifugal pump. The presence of 4% entrained vapor in a fluid can cause a reduction in pump capacity in excess of 40%.

305.5.2 Submergence of the suction pipe on a

horizontal or vertical pump is critical to proper performance. Insufficient submergence can induce vortexing or other problems that can affect pump performance.

305.5.3 The pump and the system involved in the performance test should be checked for leakage. Excessive leakage can cause erroneous performance indications. The piping should also be checked to determine if there are any bypass lines that may divert flow from the flow-measuring device. It should be noted that a line with an apparently closed valve could still be diverting fluid from the flow-measuring device if there is internal damage to the valve.

305.5.4 The pump stuffing box should be checked. Tight stuffing box packing can cause erroneous power reading on smaller pumps. The same condition can exist with improperly adjusted mechanical seals. Turning the pump over by hand prior to starting can give an indication of such a problem.

305.6 *Internal Mechanical Condition:* The wear surface clearances, impeller adjustments, and general internal wear can affect performance. The trouble-shooting section should be used to diagnose problems.

306.0 *Test Plan*

A test plan should be prepared taking into account the foregoing considerations and the conditions of the specific system and equipment.

400.0 INSTRUMENTATION AND METHODS OF MEASUREMENTS

401.0 *Introduction*

This section presents detailed information on instruments and methods of measurement most commonly used in the field testing of centrifugal pumps.

401.1 For pump testing, it is necessary to determine:

401.1.1 Pressure (suction, discharge, barometric),

401.1.2 Temperature (liquid, and local ambient air),

401.1.3 Rate of flow,

401.1.4 Pump speed,

401.1.5 Driver power input.

401.2 For the liquid being pumped:

401.2.1 Specific gravity,

401.2.2 Viscosity,

401.2.3 Vapor pressure.

402.0 *Accuracy and Precision*

402.1 *Accuracy*

Instruments or measuring devices should be checked over the expected operating range against known standards. The accuracy of all data should be determined, and the effects on the test results considered.

402.2 *Precision*

The precision of an instrument is usually reflected by its smallest scale graduations. Instrument readings recorded should reflect this precision.

403.0 *Measurement of Test Parameters*

403.1 *Instruments and Indicating Devices*

Instruments and indicating devices suitable for this test are given in the following.

403.2 *Pressure Measurements*

403.2.1 The measurement of pressure for the determination of head is carried out by pressure-indicating devices (columns, manometers, gauges or transducers) connected with the liquid passage through pressure taps or, when required, through pressure transmitters which are in turn connected to the liquid passage through pressure taps. When transmitters are used, the transmitter and the pressure-indicating device are considered as one complete instrument and are to be calibrated as a unit.

403.2.2 All gauge line connections must be tight. All instrument hose, piping, fittings, valves, etc., should be checked under pressure prior to the test to assure that there are no leaks.

403.2.3 The suction head could be approximated by measuring the distance the level of the suction liquid is above or below the centerline of the pump suction minus the calculated losses in the suction piping plus the suction surface pressure.

403.2.4 *Pressure Taps*

403.2.4.1 Pressure taps in the pipe should be flush with and normal to the wall of a cylindrical portion of the liquid passage. The interior edge of the pressure tap hole should be free of burrs and roughness as these will cause turbulence which may interfere with obtaining a true pressure reading.

403.2.4.2 Pressure readings may vary at different points on the periphery of a pipe, owing to disturbances in the flow pattern caused by internal roughness, a close proximity to fittings, a misplaced gasket, etc. Therefore, the pressure should preferably be determined at a location where flow disturbances are expected to be minimal.

403.2.4.3 Many materials can cause difficulties or inaccurate readings by corroding, plugging and freezing or condensing in these pressure connections at atmospheric temperatures or operating pressures. To prevent errors it may be necessary to use a buffer fluid between the measuring device and the operating fluid. Measuring devices may also be isolated by diaphragms.

403.3 *Temperature Measurement*

403.3.1 Temperature may be measured by using one of the measuring devices listed below:

403.3.1.1 Liquid-in-glass thermometer,

403.3.1.2 Bimetallic dial thermometer,

403.3.1.3 Thermocouples,

403.3.1.4 Resistance Temperature Device (R.T.D.).

403.3.2 The liquid-in-glass and bimetallic dial thermometers are the least expensive and are direct reading. The thermocouple and resistance temperature device require a potentiometric instrument for readout. The selection of which measuring device to use depends upon accuracy, cost, installation, and safety considerations.

403.3.3 Generally temperature variations are not crucial to test results. If temperature variations affect fluid characteristics, special considerations may be required.

403.4 *Capacity Measurement*

403.4.1 The measurement of capacity may be carried out by rate-of-flow meters, weighing, or volume methods. The reliability as well as the accuracy of weight or volume methods makes them preferable to rate-of-flow meters because the calibration of many rate-of-flow meters varies with viscosity.

403.4.2 By Weight: The measurement of capacity by weight depends upon the accuracy of the scale used and the measurement time. This method involves determining the change in the gross weight of a vessel on the suction or discharge of the pump in a given period of time.

403.4.3 By Volume: This method involves determining the volume of liquid being transferred to or from a vessel of known dimensions over a given period of time. Volume may be determined by measurement of the change in liquid level in a vessel. In this case, the liquid should be introduced into the tank with a diverter, and suitable baffles should be used in the tank to reduce swirls and eliminate surface effects.

403.4.4 By Venturi Meters, Orifice Meters, and Nozzles: Venturi meters, square-edged concentric orifice plates, and circular nozzles of the convergent type (both the submerged-flow and the free-discharge type) are acceptable instruments for field testing of centrifugal pumps. The installation and sizing of these meters should be in accordance with the manufacturer's recommendations. Selection of these meters should be based on accuracy, cost, installation, and safety considerations.

403.4.5 By Other Instruments: Rotameters, displacement meters, Doppler-effect meters and magnetic flowmeters, and Pitot tubes may be used. However, they should be calibrated by a weight or volume technique using the test liquid at conditions of rate of flow and temperature to be encountered in the test. The instrument manufacturer should be consulted for test liquid effects.

403.5 *Speed Measurement:* The pump speed should be measured by any reliable revolution counter or a speed-measuring device, such as stroboscope, hand-held tachometer, electronic counter with pulse generator or vibrating-reed tachometer.

403.6 *Power Measurement*

403.6.1 The majority of centrifugal pumps in the field are driven by a.c. electrical motors. The power input to a direct-connected pump can be accurately measured by using the electrical power in watts and multiplying by the efficiency of the motor to determine the power input to the pump shaft.

403.6.2 When the overall efficiency of a pump and driver is to be determined, this procedure shall apply to the determination of the pump performance only. The appropriate test procedure for the driver should be referred to in determining driver input.

403.6.3 The power input to the motor can be measured at the motor terminals by any one of the following acceptable methods:

403.6.3.1 Polyphase wattmeter,

403.6.3.2 Single-phase wattmeter,

403.6.3.3 Voltmeter, ammeter and power factor where required.

403.6.4 Power input from other drivers must be obtained in cooperation with the manufacturers of such equipment.

500.0 TEST PROCEDURES

501.0 *Pre-Run Data*

The pre-run data as decided by following Test Planning Section 300.0 should be obtained before the test runs are begun.

501.1 Obtain, or determine and record the properties of the liquid being pumped during the test.

501.2 Measure and record the measurements decided upon by the plan; for example, suction entry to discharge centerline on sump pumps.

501.3 Record the driver data.

501.4 Calibrate all instruments involved in the test as outlined in the appropriate portion of Section 400.0.

502.0 *Pretest Checks*

Prior to operating the pump to perform the test, the following checks should be made.

502.1 Check that the physical requirements as developed under Planning have been completed; for example: alignment.

502.2 Determine that the fluid to be pumped is available in sufficient quantities to complete all the test runs to be made.

502.3 Follow and check the piping systems to insure that fluid will be delivered to, and be discharged from the pump as required by the needs of the test. Are there branch lines to be accounted for in obtaining flow, etc.?

502.4 Check the driver and pump for direction of rotation, freedom of turning and coupling alignment.

502.5 Check the instrumentation to be sure sensing systems are full and are not influenced by ambient conditions.

503.0 *Trial Run Checks*

503.1 Open the pump suction and discharge valves. In some cases it may be desirable to begin with a partially open discharge valve to avoid the effect of full flow surge on some system components or to reduce power requirements at start-up. Make sure the pump is primed.

503.2 Start the pump by applying power or admitting steam to the driver.

503.3 Observe this initial operation check for vibration, shaft seal leakage, pump priming or need for venting and heat build-up in the shaft seal, bearings, or pump and motor casings. The discharge pressure should be appropriate for the conditions.

503.4 Check to determine the pump is delivering a flow which is consistent with that expected.

503.5 Use a stroboscope, a direct-reading tachometer, or other device selected during the test planning to determine if the design speed is attained.

503.6 When pressures, flow rates, speed, temperatures, and power readings are stable or within acceptable limits of fluctuation, proceed with the test run or runs necessary for the desired determinations.

504.0 *Test Run*

The data enumerated in this section will establish all the operating characteristics, but the data to be collected during a run may be abbreviated as decided in the test plan. Obtaining this bench-mark information on a new pump or a newly overhauled pump will provide the basis for evaluation of any future tests: any deficiencies observed in the trial run must be corrected, or if none are observed, the test run can proceed.

504.1 Data required are suction pressure, discharge pressure, absolute pressure on the surface of the liquid being pumped, liquid temperature at the suction, flow rate, power readings, and pump speed.

504.2 Line up and start the test systems as stated under the Trial Run Checks, Section 503.0, or proceed if the system is already operating.

504.3 Establish the flow desired and allow conditions to stablize.

504.4 Take the required readings and record on a data sheet, such as the example in the Appendix, Figure 802.2.1. The need for simultaneous data depends on whether readings are fluctuating. Obviously, if there is no change, one person can proceed from data point to data point and collect the readings.

504.5 Change the flow rate, then return to the first data point and allow time for the conditions to stablilize.

504.6 If taking data to establish characteristic curves, the data points for increasing flow rates should be repeated with decreasing flows.

504.7 Check the pump, instrumentation, and piping system after the test is completed to see that no changes have occurred which would cause erroneous readings.

600.0 COMPUTATION OF RESULTS

601.0 *Results Required*

The usual results desired are the computed coordinates based on the test data which provide a comparison with the characteristic curves issued by the pump manufacturer. (See Figure 702.2.1, Characteristic Curve.)

601.1 The coordinates needed for a point on the characteristic curves are total head and capacity, power and capacity, available net positive suction head and capacity, or efficiency and capacity.

602.0 *Total Developed Head*

Total developed head is the hydraulic energy added to the pumped fluid, i.e., the algebraic difference between the total discharge head and total suction head.

602.1 Readings of pressure gauges in customary (U.S.) units are converted to pounds per square inch and then to feet of liquid head to obtain static head. Readings of SI Units are converted to pounds per square inch and then to feet of head of liquid. (See example in Section 802.3.) Static head is defined in Section 200.0. Due regard must be paid to correcting for gauge location with regard to the chosen datum line.

602.2 The velocity used for determining the velocity

head (suction or discharge), as defined in Section 200.0, may be obtained from the volume rate of flow and the cross-sectional area at the measurement section. Velocity in ft/sec is determined by:

$$v = \frac{Q}{A}$$

Q = volume flow rate, ft^3/sec,

A = cross-sectional area, ft^2.

602.3 Velocity to be used for determining flow may also be calculated from the differential head across a Venturi, orifice, or nozzle throat:

$$v = C\ 2gh,$$

v = velocity, ft/sec

C = meter coefficient

h = differential head of fluid pumped as computed from the manometer head, ft

g = acceleration of gravity, 32.174 ft/sec^2.

603.0 *Capacity*

Capacity or volume rate of flow is in terms of cubic feet per second for calculation purposes and as cubic meter per second or gallons per minute for reporting purposes. The following equations may be used as they apply:

$$Q,\ ft^3/sec = \frac{\text{weight rate of flow, lb/sec}}{\text{specific weight, lb/ft}^3}$$

$$Q,\ ft^3/sec = 35\ \frac{\text{weight rate of flow, kg/sec}}{\text{specific weight, kg/meter}^3}$$

$$Q,\ ft^3/sec = \frac{\text{volume rate of flow, gal/min}}{(60\ \text{sec/min})\ (7.48\ \text{gal/ft}^3)}$$

$$Q,\ ft^3/sec = (\text{velocity, ft/sec})\ (\text{area, ft}^2).$$

604.0 *Power*

The characteristic power curve is an indication of the rate of doing work applied to the pump.

604.1 Driver Horsepower or

$$Hp = \frac{\text{(watts input)(motor efficiency)}}{746} = \text{horsepower}$$

Where a wattmeter is not used to obtain the input, the following is used on 3-phase motors:

watts input = 1.73 × phase voltage
× average phase amperage
× power factor.

604.2 Where other machines provide the power input, the manufacturer can supply means of obtaining this power. For example, a steam turbine driver might use nozzle block steam pressure, exhaust steam pressure and efficiency to determine the brake horsepower.

605.0 *Efficiency*

The pump power output combined with the efficiency (P_W/H_P) at any point may provide a check on the driver brake horsepower, or combined with the driver brake horsepower, used to determine efficiency.

605.1 Pump hydraulic power output is found by the following formula:

$$P_w = \frac{\text{lb liquid pumped/min} \times \text{total developed head*}}{33{,}000}$$

If the liquid is of specific gravity (s) with reference to water at 69 °F,

$$P_w = \frac{s\ Q\,(gpm)\ H}{3960}$$

in US Customary units or in metric units

$$P_w = Q\,(p_1 - p_2)g$$

P_w = hydraulic power, watts

Q = volumetric flow rate, m^3/sec

p_1 = discharge pressure, *Pa*

p_2 = suction pressure, *Pa.*

g = 9.807 m/sec^2

605.2 Pump efficiency is then determined as follows:

$$Eff = \frac{\text{output}}{\text{input}} = \frac{P_w}{Hp}$$

606.0 *Speed*

Where the test speed is different than the speed used to develop the characteristic curves, the values of the coordinates may be estimated by the following equations:

$$Q = Q_t\left(\frac{n}{n_t}\right)$$

$$H = H\left(\frac{n}{n_t}\right)^2$$

$$NPSH = NPSH_t\left(\frac{n}{n_t}\right)^2$$

$$P_w = P_{wt}\left(\frac{n}{n_t}\right)^3$$

where the sucscript "t" refers to the conditions at the test speed. Variations between the test speed and the characteristic curve speed that exceed 10% could affect the validity of the comparison.

---

*Not discharge pressure.

Sample Test-Run Data Sheet

Pump Type ____________	Pump Seria! No. ____________
Pump Size ____________	Rated Capactiy ____________ gpm
Pump Manufacturer ____________	Rated Head ____________ ft
Pump No. of Stages ____________	Impeller Diameter ____________
Driver Type ____________	Rated Horsepower ____________ hp
Driver Speed ____________ rpm	Driver Serial No. ____________
Suction Temperature ____________ °F	Suction Pressure ____________ psig
Liquid Type ____________	
Specific Gravity ____________	Viscosity Absolute ____________ cp
Barometric Pressure ____________ mm.Hg	Pressure (liquid surface) ____________ psig
Discharge Pressure ____________ psig	
Motor Input Amps ____________ amps	Motor Input Volts ____________ volts
Motor Power Factor ____________	Motor Power Input ____________ watts
Turbine Nozzle Ring Pressure ____________ psig	No. of Nozzles ____________
Turbine Back Pressure ____________ psig	

System Sketch:

700.0 INTERPRETATION OF RESULTS

701.0 *Introduction*

The graphical presentation of data provides the most effective method of explaining results. Each curve should be identified and dated with the pertinent information such as pump serial number, impeller designation, impeller diameter, speed, and liquid pumped. Field test results are generated for comparison against the manufacturer's published performance. The published performance or, preferably, the results of an actual factory test, should be used when available.

702.0 *Discussion*

702.1 Data obtained under field conditions may not represent a smooth curve. The curve represented by the data will be an average. If the data points deviate significantly from the average curve, investigation of the reasons for these deviations should be made.

702.2 There is a given relationship between horsepower, total developed head, capacity, and $NPSH_R$ requirements. As a cross-check on the field test results, one parameter can be compared to the others using the manufacturer's published characteristic curves. If the head and horsepower are known, the capacity can be estimated. If the $NPSH_R$ is known, the capacity at that point can be estimated. If the horsepower is known, the capacity can be estimated. See the Characteristic Curve (Figure 702.2.1) for the relationship between these parameters.

702.3 Capacity measurements are usually the most difficult to obtain. In normal field installations you will be unlikely to find instrumentation for measurement of capacity. Indirect determination of capacity may be necessary. The use of the performance curve similar to that in Figure 702.2.1 is an acceptable method of indirect determination.

702.4 The head measurement is the easiest to accomplish in the field. A common method of checking pump performance is to determine the pressure at zero capacity or shutoff. Caution should be used to operate pumps above the manufacturer's minimum flow values. This removes one of the variables in the relationship and allows you to determine proper impeller diameter or speed. Shut-off head is a function of diameter and speed. Speed or diameter can be verified by this initial check. Caution should be used on turbine and propeller-type pumps where the shut-off horsepower may exceed the motor rating due to a rising horsepower curve.

702.5 Various methods of field performance checks can be conducted depending on the shape of the performance curve for the particular pump being monitored. The following curves represent varying types of centrifugal pump performance. In Figure 702.5.1, if the head capacity curve is rather shallow as shown by the solid line, measurement of capacity by monitoring pressure would not be effective as a large change in capacity represents a very small change in pressure. The dotted line represents a very steep head-capacity curve. Under these conditions, a measurement of pressure can give fairly accurate measurement of capacity.

TYPICAL PUMP PERFORMANCE CURVE

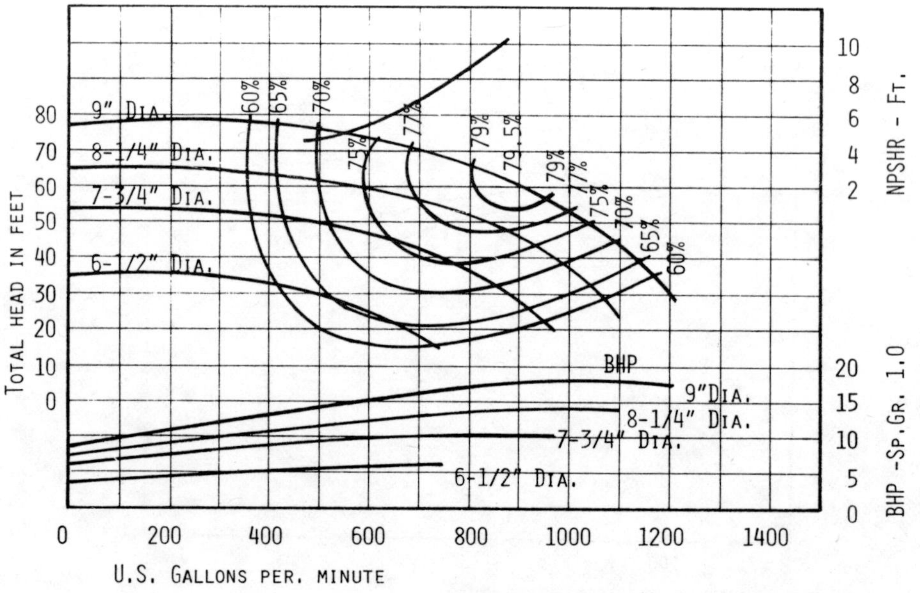

Figure 702.2.1

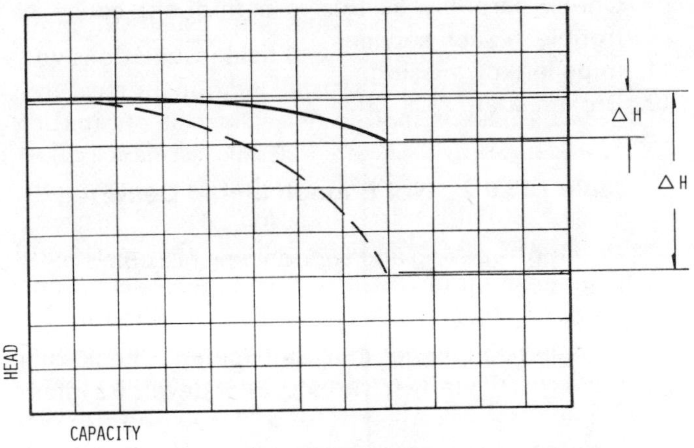

Figure 702.5.1

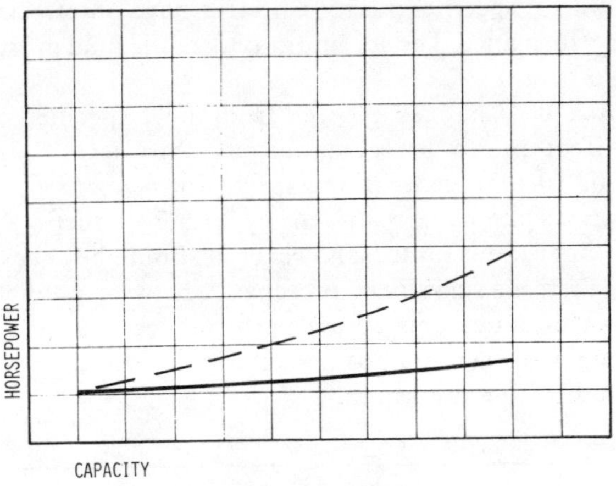

Figure 702.5.2

702.6 Horsepower versus capacity can also be used to estimate capacity. The flat horsepower curve, Figure 702.5.2, represented by the solid line, would not permit such determination. Other methods would have to be used. The dotted line represents a very steep horsepower curve and would lend itself to indirect capacity determination.

702.7 The same approach can be used with net positive suction head requirements. If the net positive suction head available can be varied, a pump within a steep curve as shown by the dotted line could be monitored for capacity by varying the $NPSH_A$ available until the pump total head drops. If the $NPSH_R$ line is shallow, indirect measurement of capacity using this method is not practical.

702.8 It should be noted that viscosity will affect performance of a centrifugal pump. The curve (Figure 702.8.1) taken from the Hydraulic Institute Standards shows the effect of viscosity on the performance of a centrifugal pump. Viscosity characteristics of the fluid that is being pumped should be determined. Predicted divergence from the manufacturer's published water performance should be determined.

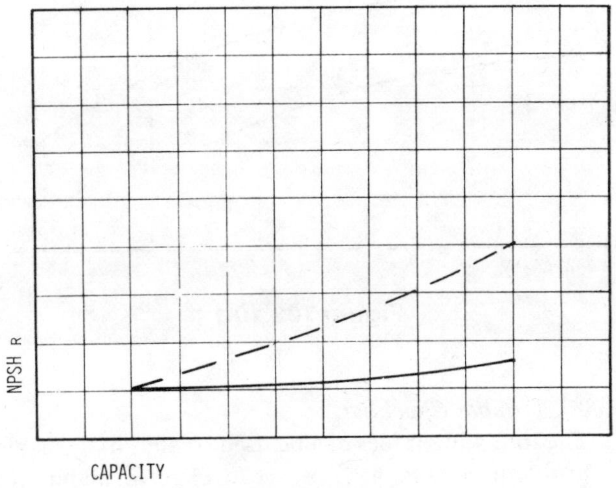

Figure 702.7.1

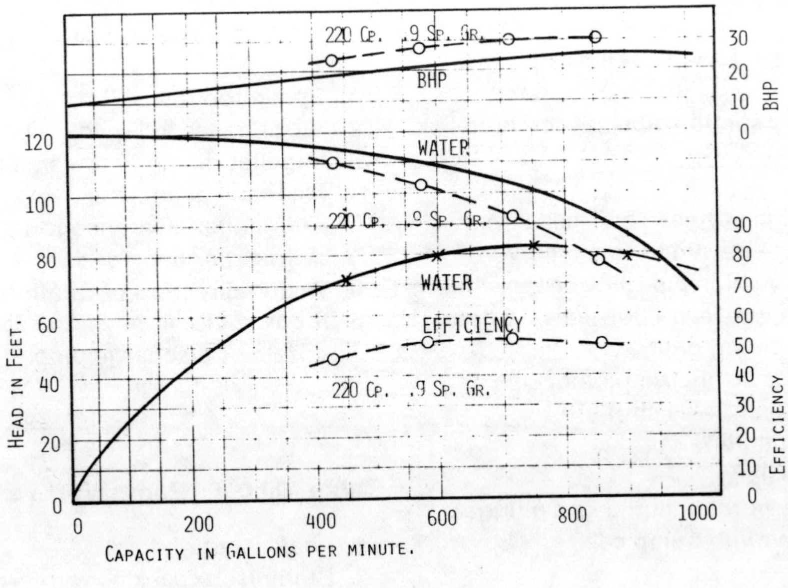

Figure 702.8.1

702.9 Specific gravity will also have an effect on performance. The specific gravity of the fluid must be known.

702.10 Changes in speed from those specified can affect pump performance, as shown in Figure 702.10.1. In general, the speed will be as predicted when pumps are driven by either induction or synchronous motors; however, it should be checked if there is a significant discrepancy between predicted performance and actual performance. Turbine or engine-driven pumps may vary in speed. Speed verification is essential.

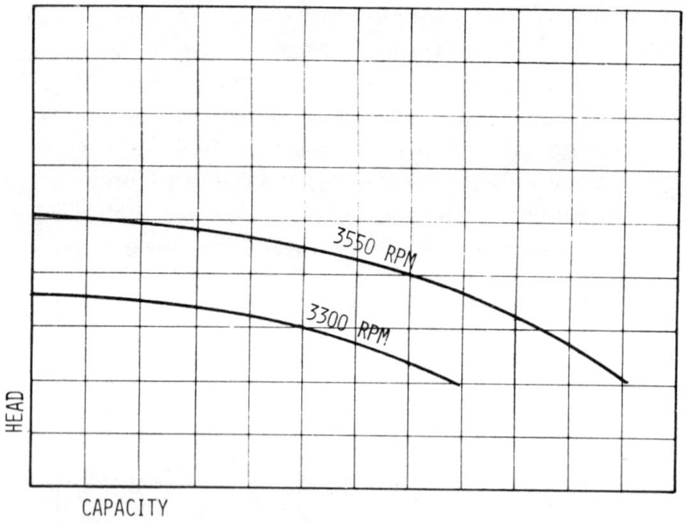

Figure 702.10.1

703.0 *Trouble Checklist*

The following tables can be used to identify possible problem areas that can affect performance in centrifugal pumps.

Table 703.0.1 No Liquid Delivered

1. Priming — casing and suction pipe not completely filled with liquid.
2. Speed too low.
3. System head too high. Check total head (particularly friction loss) or develop system curve.
4. Suction lift too high (suction pipe may be too small or long, causing excessive friction loss). Check with vacuum or compound gauge.
5. Impeller or suction or discharge pipe or opening completely plugged (i.e., valve shut off).
6. Wrong direction of rotation.
7. Air pocket in suction line.
8. Stuffing box packing worn, or liquid seal plugged, allowing leakage of air into pump casing.
9. Air leak in suction line.
10. Not enough suction head for hot or volatile liquids. Check carefully as this is a frequent cause of trouble in such service.
11. Impeller key missing.
12. Broken shaft.

Table 703.0.2 Not Enough Liquid Delivered

1. Priming — casing and suction pipe not completely filled with liquid.
2. Speed too low.
3. System head higher than anticipated. Check total head (particularly friction loss) or develop system curve.
4. Suction lift too high (suction pipe may be too small or long, causing excessive friction loss). Check with vacuum or compound gauge.
5. Impeller suction or discharge opening partially plugged.
6. Wrong direction rotation.
7. Air pocket in suction line.
8. Stuffing box packing worn, or liquid seal plugged, allowing leakage of air into pump casing.
9. Air leak in suction line.
10. Not enough suction head for hot or volatile liquids. Check carefully as this is a frequent cause of trouble in such service.
11. Foot valve too small.
12. Foot valve or suction pipe not immersed deeply enough.
13. Mechanical defects:
    a. Impeller clearance too great.
    b. Impeller damaged.
    c. Wrong impeller (i.e., size).
14. Vortexing.
15. Entrained gases in liquid.
16. Impeller key missing.

Table 703.0.3 Not Enough Pressure

1. Speed too low.
2. Air or gases in liquid.
3. Impeller diameter too small.
4. Mechanical defects:
    a. Impeller clearance too great.
    b. Impeller damaged.
5. Wrong direction of rotation.
6. Incorrect location of pressure gauge on discharge nozzle or discharge pipe.

Table 703.0.4 Pump Works a While and Then Quits

1. Leaky suction line.
2. Stuffing box packing worn—or liquid seal plugged—allowing leakage of air into pump casing.

3. Mechanical defects:
   a. Broken shaft.
   b. Broken coupling.
4. Moto stops due to overload trip out.
5. Vortexing or insufficient $NPSH_A$.

Table 703.0.5 Noise

1. Insufficient $NPSH_A$.
2. Bearings.
3. Foreign object in pump casings.
4. Vortexing.

703.1 *Cavitation Characteristic*
How does one determine that a given pump is short of $NPSH_A$ and is cavitating? The distinctive crackling noise of cavitation is often drowned out by the ambient noise of motors and other machinery.
703.2 *Cavitation Phenomenon*
Cavitation occurs when the absolute pressure within the pump impeller passages falls below the vapor pressure of the liquid at that point and bubbles of vapor are formed. These bubbles subsequently collapse as they move into higher pressure regions.

703.2.1 A pump that is cavitating will exhibit a distinctive sharp crackling sound, vibration, loss of efficiency and head at a given capacity.

703.2.2 The sudden collapsing of the vapor bubbles will cause pitting and erosion of the impeller material in areas where these bubbles collapse adjacent to the metal surfaces. Severe damage can occur to the impeller in a relatively short time. Also, the mechanical seal can be severely damaged.

703.2.3 True cavitation requires actual vaporization and condensation of the liquid being pumped. Other phenomena, such as dissolved gases coming out of solution and separation of the liquid from the metal surfaces at a point where the surface is convex, are not cavitation in the strict sense. Although these phenomena can cause turbulence, noise and vibration, they will not cause the typical pitting and erosion of the metal parts of the pump.

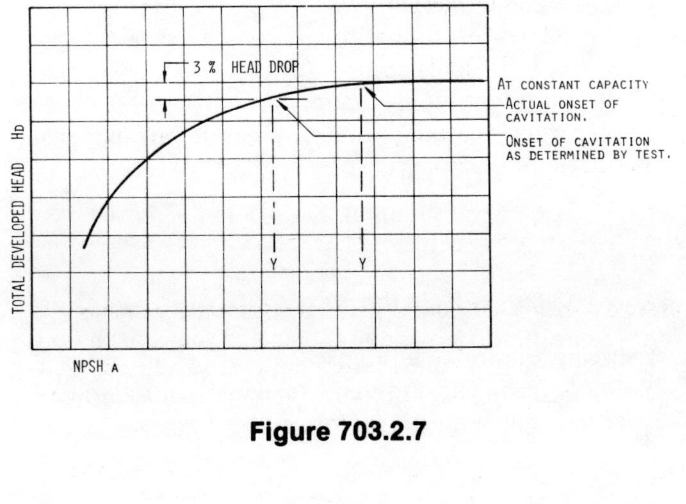

Figure 703.2.7

703.2.4 It bears repeating that a centrifugal pump will begin to cavitate when the suction head is not sufficient to maintain local pressures above the vapor pressure through the flow passages. Obviously, the parts most vulnerable to damage are the lowest pressure parts of the pump such as the rear side of the impeller vanes near the suction eye and inlet edge, also the front shroud on a closed impeller where the liquid turns from axial to radial direction. The back shroud of open impellers with pump-out vanes and exit tips of impeller vanes on pumps operating to the far right of the performance curve are also subject to cavitation damage.

703.2.5 Pump casings can also show cavitation damage on the low pressure side of the "cut water" or tongue in volute pumps and on the low pressure side of the diffuser vanes on diffuser-type pumps.

703.2.6 A practical method would be to allow the pump to operate at a constant capacity where Net Positive Suction Head Available, $NPSH_A$, is known to be sufficient, then gradually reduce $NPSH_A$ by either throttling the suction thereby decreasing h_s or lowering its pressure by decreasing h_{ps}. Changing any one of these factors will cause a decrease in $NPSH_A$; it is best to modify only one factor as it obviously makes for easier calculations.

703.2.7 As $NPSH_A$ is decreased, the discharge head developed by the pump will remain essentially constant until the point is reached where Net Positive Suction Head Available, $NPSH_A$, equals Net Positive Suction Head Required, $NPSH_R$, and cavitation begins. At this point, and before any distinctive noise is heard, the discharge head will show a drop. It is quite difficult to determine the exact point at which cavitation starts, but based on the sensitivity of the discharge head indication, it is possible to detect a drop in head of 2% or 3%. The onset of cavitation ($NPSH_R$) has now been determined with reasonable accuracy for a given flow differing only from the actual by some slight value.

703.2.8 Changing the pump capacity to a different constant value and repeating the process will determine a second point. By this means an $NPSH_R$ curve can be drawn for that particular pump in that particular service.

703.3 *Other Liquids*

703.3.1 To avoid cavitation the $NPSH_A$ value should not be reduced below the $NPSH_R$ value. A reduction in $NPSH_R$ below that shown for cold water is possible. Pumps handling hot water or certain hydrocarbons can operate safely with less $NPSH_A$ than would be required for cold water. The magnitude of the $NPSH_R$ reduction is a function of the pumping temperature and vapor pressure. This phenomenon would tend to explain why a given pump would seem to be operating satisfactorily even though it is requiring more than the calculated $NPSH_A$.

703.3.2 Hydraulic Institute published a curve to show just how much $NPSH_R$ reduction a certain hydrocarbon will allow. Care should be used in applying this curve if the liquid being pumped is not one of those shown on the curve.

703.4 *Reduction and Control of Cavitation Damage*

If a given pump has suffered damage from cavitation and it is not possible to increase the $NPSH_A$ or reduce $NPSH_R$ by means of operating at a lower flow, certain things can be done to alleviate or reduce continued damage short of massive modifications such as lowering the pump.

703.4.1 Certain materials have relatively great resistance to cavitation attack. Cast iron and other soft materials are especially prone to damage. It may be possible to replace pump impeller and/or casing with a material which has a much higher resistance to damage.

703.4.2 The leading edge of the impeller vanes can be cut back and perhaps the impeller eye enlarged thereby reducing velocity and its attendant pressure drop. The pump manufacturer should be contacted concerning an alternate impeller pattern that includes these modifications that may still perform with satisfactory head, capacity, and efficiency.

703.4.3 The existing impeller vanes may be sharpened by filing to a narrow leading edge.

703.4.4 Pre-rotation in the suction piping can be reduced by installation of a simple straightening vane. Care should be taken to insert vanes with a minimum pressure drop.

703.4.5 Impeller inlet vanes and casing can be coated with a resilient, nonmetallic coating material that provides a cushioning effect to the shock waves produced by the collapsing bubbles. Applying such coatings is a special procedure, will probably affect the head/capacity performance of the pump and may not be possible on high-speed units.

703.4.6 For materials near their boiling point the suction line may be cooled to raise the $NPSH_A$.

800.0 APPENDIX

801.0 *Nomenclature*

Term	Symbol	Units: Customary (U.S.)	Units: S.I.*
Pump Total Head	H	ft of liquid	kPa
Total Discharge Head	h_d	ft of liquid	kPa
Total Suction Head	h_s	ft of liquid	kPa
Total Suction Lift	h_{gs}	ft of liquid	kPa
Static Discharge Head	h_{sd}	ft of liquid	kPa
Discharge Velocity Head	h_{vd}	ft of liquid	kPa
Static Suction Head	h_{ss}	ft of liquid	kPa
Suction Surface Pressure	h_{ps}	ft of liquid	kPa
Suction Friction Head	h_{fs}	ft of liquid	kPa
Suction Velocity Head	h_{vs}	ft of liquid	kPa
Datum Correction Suction	Z_s	ft of liquid	kPa
Datum Correction Discharge	Z_d	ft of liquid	kPa
Liquid Vapor Pressure	h_{vpa}	psia	kPa
Local Gravitational Acceleration	g	ft/sec^2	m/sec^2
Suction Surface Pressure Absolute	h_{psa}	psia	kPa
Pressure Head	h_p	ft of liquid	kPa
Gauge Pressure	P_g	lb/in^2	kPa

*Note: To convert customary units to SI or the reverse, you must convert feet of liquid to pounds per square inch then to kilopascals and to meters, or meters to kilopascals to pounds per square inch to feet. As an example: a pump discharge gauge reads 200 kPa and you wish to convert that reading to head in feet. You must first convert kPa to psi (200 kPa × 0.0145 = 2.9 psi), then convert psi to head in feet (psi × 2.31/specific gravity = head in feet), for specific gravity of 1.1 (2.9 × 2.31/1.1 = 6.1 feet of liquid).

Term	Symbol	Units Customary (U.S.)	S.I.
Capacity of Pump	Q	gpm	m^3/sec
Impeller Diameter	D	inches	m
Velocity	v	ft/sec	m/sec
Velocity Head	h_v	ft of liquid	kPa
Specific Gravity	s	Dimensionless	
Specific Weight (Mass)	γ	lb/ft^3	kg/m^3
Power	Hp	horsepower	kW
Temperature	T	°F	°C
Speed of Rotation	n	rpm	rad/s
Length	L	ft	m
Net Positive Suction Head Available	$NPSH_A$	ft of liquid	kPa
Net Positive Suction Head Required	$NPSH_R$	ft of liquid	kPa
Pump Efficiency	Eff	percent	%
Power Factor	Pf	Dimensionless	
Current	I	amps	A
Voltage	E	volts	V
Hydraulic Power	P_w	horsepower	kW
Discharge Pressure	P_2	psig	kPa
Suction Pressure	P_1	psig	kPa
Viscosity	μ	centipoise	Pa·s

802.0 *Sample Test Results*

The example problem is shown to demonstrate calculations needed to establish various pump characteristics at a single operating point. When complete performance curves are required, several points must be determined.

802.1 *System Configuration* See Figure 802.1

802.2 *Sample Data Log Sheet*

Figure 802.2.1 is an example of a log sheet used to assemble data which will be used in the sample calculation. Sample calculation is based on only one flow condition (1000 gpm). All other flow conditions will be taken from Figure 802.2.2.

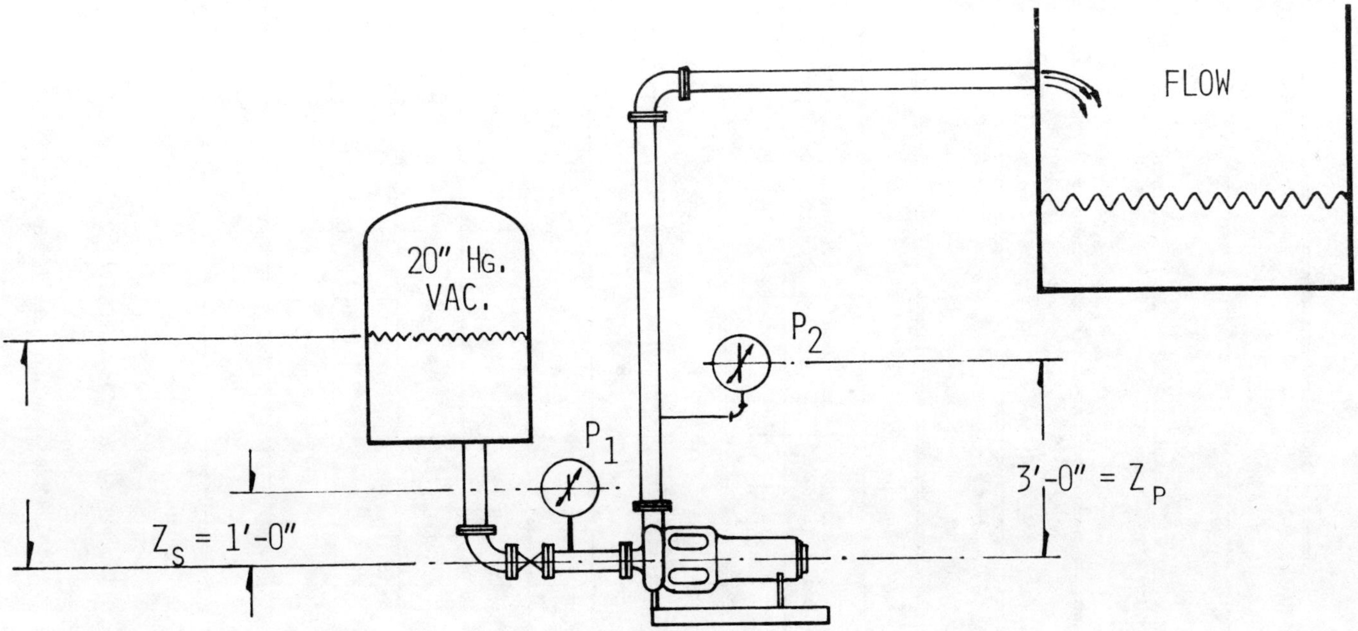

Figure 802.1

RECORD OF PUMP TEST

Mark each pump with test number

MANUFACTURER Brand X	DISCARGE DIAMETER 3″	SUCTION DIAMETER 4″	IMPELLER DIAMETER 12″	TEST NO. 1
MODEL DESCRIPTION	LIQUID TEMP. 100 °F	BAROMETRIC PRESSURE 14.7	SPECIFIC GRAVITY 0.9944	DATE 11-17-80

MANUFACTURER SERIAL NO.	G.P.M. AT 1000	FT. TOTAL HEAD 98	VAPOR PRESSURE 0.947	VISCOSITY 1 cp	TESTED BY

MOTOR (DRIVER) 4.0 HP MOTOR	MOTOR SPEED 1750 RPM		LINE VOLTAGE 460 VOLTS

EFFICIENCY % 90	POWER FACTOR 0.875	CYCLES/HERTZ 60	PHASE 3	

RATE OF FLOW	P_1 SUCTION HEAD GA.	P_2 DISCHARGE HEAD GA.	Z_d DISCHARGE GA. CORRECTION	Z_s SUCTION GA. CORRECTION	TOTAL HEAD	POWER INPUT	BRAKE H.P.	PUMP SPEED	CORRECTED FLOW	CORRECTED HEAD	CORRECTED B.H.P.	WATER H.P.	PUMP EFF.
GPM	PSIA	PSIA	FEET	FEET	FEET	AMPS	HP	RPM	GPH	FEET	HP	WHP	%
0	11	68	3.0	1.0	135	18	15	1750	None	None	None	0	0
500	10.5	63	↓	↓	125	26.2	21.6	1745				15.8	73
800	10	57			112	31	27.6	1749				22.6	82
1000*	9.0	49			96	36	30	1750				24.9	83
1100	8.5	46			90	37	30.8	1747				25	81
1200	8	42			80	37.3	31.1	1752				24.3	78
1400	7	30			55	39	32.4	1750				19.4	60
1500	6.3	22	↓	↓	40	41.5	34.4	1753				15.1	44

REMARKS: Piping:
Suction — 6″ SCH 40 STEEL
Discharge — 6″ SCH 40 STEEL

*Condition used in sample calculation, Sec. 802.3

Figure 802.2.1

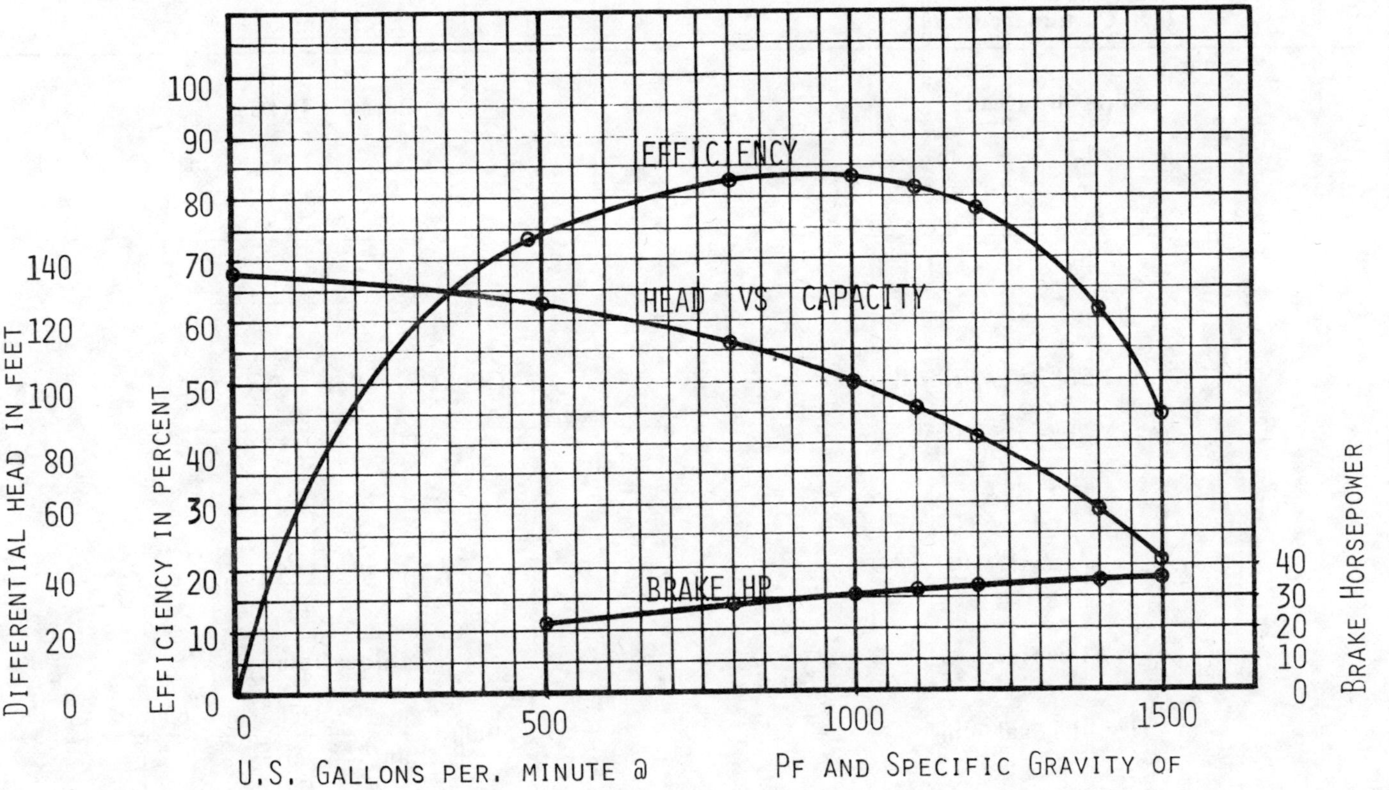

Figure 802.2.2

802.3 *Flow Measurement*
Flow was determined by measuring the difference in level in the discharge tank over a time period.

802.4 *Power for Sample Problem*
The power factor shown in the data was obtained from the motor manufacturer's typical values for the type and size motor. The motor efficiency was obtained in the same manner.

U.S. Customary Units

Pressure Head

$$h_p = \frac{P_g \frac{lb_f}{ft^2} \quad 32.2 \frac{lb_m\ ft}{lb_f\ sec^2}}{\gamma \frac{lb_m}{ft^3} \quad g \frac{ft}{sec^2}}$$

$$(144)\ P_g\ (PSIA) = P_g\ (lb_f/ft^2)$$

$$(62.4)\ s = \gamma\ (lb_m/ft^3)$$

$$g = 32.2\ (ft/sec^2)$$

$$h_p = 2.31 \frac{P_g(PSIA)}{s}\ ft$$

Flow Rate

$$Q = 1000\ gal/min$$

Static Suction Head, h_{ss}

$$P_1 = 9\ PSIA$$

$$h_{ss} = 2.31 \frac{P_l\ (PSIA)}{s}\ ft$$

$$s = 0.9944$$

$$h_{ss} = 2.31\ (9)\ /\ (0.9944)$$

$$= 20.91\ feet$$

Static Discharge Head, h_{sd}

$$P_2 = 49.3\ PSIA$$

$$h_{sd} = 2.31 \frac{P_2\ (PSIA)}{s}\ ft$$

$$h_{sd} = 2.31\ (49.3)\ /\ (0.9944)$$

$$= 114.52\ ft$$

S.I. Units

Pressure Head

$$h_p = \frac{P_g \frac{N}{m^2} \quad 1 \frac{kg\ \ m}{N\ sec^2}}{\gamma \frac{kg}{m^3} \quad g \frac{m}{sec^2}}$$

$$(1000)\ P_g\ (kPa) = P_g\ (N/m^2)$$

$$(1000)\ s = \gamma\ (kg/m^3)$$

$$g = 9.807\ (m/sec^2)$$

$$h_p = 0.102 \frac{P_g\ (kPa)}{s}\ meters$$

Flow Rate

$$Q = 1000 \frac{gal}{min} \frac{0.003785\ m^3}{gal} \frac{60\ min}{hr}$$

$$= 227.1\ m^3/hr$$

Static Suction Head, h_{ss}

$$P_1 = 9\ PSIA \frac{6.894\ kPa}{PSIA}$$

$$= 62.05\ kPa$$

$$h_{ss} = 0.102 \frac{P_l(kPa)}{s}\ meters$$

$$s = 0.9944$$

$$h_{ss} = 0.102\ (62.05)\ /\ (0.9944)$$

$$= 6.36\ meters$$

Static Discharge Head, h_{sd}

$$P_2 = 49.3\ PSIA \frac{6.894\ kPa}{PSIA}$$

$$= 339.87\ kPa$$

$$h_{sd} = 0.102 \frac{P_2\ (kPa)}{s}\ meters$$

$$h_{sd} = 0.102\ (339.87)\ /\ (0.9944)$$

$$= 34.86\ meters$$

U.S. Customary Units	S.I. Units

Suction Velocity Head, h_{vs}

$$h_{vs} = \frac{V^2}{2\,g} \text{ for 6'' line}$$

$$V = 11.1 \text{ ft/sec}$$

$$h_{vs} = \frac{(11.1 \text{ ft/sec})^2}{2\ (32.17 \text{ ft/sec}^2)}$$

$$= 1.91 \text{ ft}$$

Discharge Velocity Head, h_{vs}

$$h_{vd} = \frac{V^2}{2\,g} \text{ for 6'' line}$$

$$V = 11.1 \text{ ft/sec}$$

$$h_{vd.} = \frac{(11.1 \text{ ft/sec})^2}{2\ (32.17 \text{ ft/sec}^2)}$$

$$= 1.91 \text{ ft}$$

Datum Correction Suction, Z_s

$$Z_s = 1 \text{ ft}$$

Datum Correction Discharge, Z_d

$$Z_d = 3 \text{ ft}$$

Vapor Pressure, h_{vpa}

$$h_{vpa} = 0.947 \text{ PSIA}$$

$$h_{vp} = 2.31\ \frac{P_{vpa} \text{ (PSIA)}}{s} \text{ ft}$$

$$h_{vp} = 2.31\ (0.947)\ /\ (0.9944)$$

$$= 2.199 \text{ ft}$$

Total Suction Head, h_s

$$h_s = h_{ss} + Z_s + h_{vs}$$

$$= 20.91 + 1.0 + 1.91$$

$$= 23.82 \text{ ft}$$

S.I. Units

Suction Velocity Head, h_{vs}

$$h_{vs} = \frac{V^2}{2\,g}$$

$$V = 11.1 \text{ (ft/sec) } (0.3048 \text{ m/ft})$$

$$= 3.38 \text{ m/sec}$$

$$h_{vs} = \frac{(3.38 \text{ m/sec})^2}{2\ (9.807 \text{ m/sec}^2)}$$

$$= 0.584 \text{ m}$$

Discharge Velocity Head, h_{vs}

$$h_{vd} = \frac{V^2}{2\,g}$$

$$V = 11.1 \text{ (ft/sec) } (0.3048 \text{ m/ft})$$

$$= 3.38 \text{ m/sec}$$

$$h_{vd} = \frac{(3.38 \text{ m/sec})^2}{2\ (9.807 \text{ m/sec}^2)}$$

$$= 0.584 \text{ m}$$

Datum Correction Suction, Z_s

$$Z_s = 1 \text{ ft } (0.3048 \text{ m/ft}) = 0.3048 \text{ m}$$

Datum Correction Discharge, Z_d

$$Z_d = 3 \text{ ft } (0.3048 \text{ m/ft}) = 0.91 \text{ m}$$

Vapor Pressure, h_{vpa}

$$h_{vpa} = 0.947 \text{ PSIA } (6.894 \text{ kPa/PSIA})$$

$$= 6.529 \text{ kPa}$$

$$h_{vp} = 0.102\ \frac{P_{vpa} \text{ (kPa)}}{s} \text{ meters}$$

$$h_{vp} = 0.102\ (6.529)\ /\ 0.9944$$

$$= 0.670 \text{ m}$$

Total Suction Head, h_s

$$h_s = h_{ss} + Z_s + h_{vs}$$

$$= 6.36 + 0.3048 + 0.548$$

$$= 7.248 \text{ m}$$

U.S. Customary Units

Total Discharge Head, h_d

$$h_d = h_{sd} + Z_d + h_{vd}$$

$$= 114.52 + 3 + 1.91$$

$$= 119.43 \text{ ft}$$

Total Head, H

$$H = h_d - h_s$$

$$= 119.43 - 23.82$$

$$= 95.61 \text{ ft}$$

Driven Horsepower

$$Hp = \frac{\sqrt{3}\ E\ I\ Pf\ Eff}{746 \text{ W/hp}}$$

$$= \frac{\sqrt{3}\ (460)\ (36)\ (0.875)\ (0.90)}{746 \text{ W/hp}}$$

$$= 30.28 \text{ hp}$$

Water Horsepower

$$P_w = Q\ H\ \gamma$$

$$Q = 1000 \text{ gal/min } (\text{ft}^3/7.48\text{gal})$$

$$= 133.7 \text{ ft}^3/\text{min}$$

$$= 2.228 \text{ ft}^3/\text{sec}$$

$$H = 95.51 \text{ ft}$$

$$\gamma = 0.9944\ (62.4 \text{ lb/ft}^3)$$

$$= 62.05 \text{ lb/ft}^3$$

$$P_w = (2.228)\ (95.51)\ (62.05)$$

$$= \frac{13204 \text{ ft lb/sec}}{550 \text{ (ft lb/sec)/hp}}$$

$$= 24.01 \text{ hp}$$

S.I. Units

Total Discharge Head, h_s

$$h_d = h_{sd} + Z_d + h_{vd}$$

$$= 34.86 + 0.91 + 0.584$$

$$= 36.36 \text{ m}$$

Total Head, H

$$H = h_d - h_s$$

$$= 36.36 - 7.248$$

$$= 29.11 \text{ m}$$

Driven Horsepower

$$Hp = \sqrt{3}\ E\ I\ Pf\ Eff$$

$$= \sqrt{3}\ (460)\ (36)\ (0.875)\ (0.90)$$

$$= 22588 \text{ W} = 22.59 \text{kW}$$

Water Power

$$P_w = Q\ H\ \gamma\ g$$

$$Q = 227.1 \text{ m}^3/\text{hr } (\text{hr}/3600 \text{ sec})$$

$$= 0.06308 \text{ m}^3/\text{sec}$$

$$H = 29.10 \text{ m}$$

$$\gamma = 0.9944\ (1000 \text{ kg/m}^3)$$

$$= 994.4 \text{ kg/m}^3$$

$$g = 9.807 \text{ m/s}^2$$

$$P_w = (0.06308)\ (29.10)\ (994.4)\ (9.807)$$

$$= 17901 \text{ kg m}^2/\text{s}^3$$

$$= 17.90 \text{ kW}$$

U.S. Customary Units

Efficiency

$$Eff = \frac{P_w}{H_p} (100\%)$$

$$= \frac{24.01 \text{ hp}}{30.28 \text{ hp}} (100\%)$$

79.3%

NPSH$_A$

$$NPSH_A = h_{ss} + Z_s - h_{vp}$$

$$= 20.91 + 1 - 2.199$$

$$= 19.7 \text{ ft}$$

S.I. Units

Efficiency

$$Eff = \frac{P_w}{H_p} (100\%)$$

$$= \frac{17.90 \text{ kw}}{22.59 \text{ kw}} (100\%)$$

79.3%

NPSH$_A$

$$NPSH_A = h_{ss} + Z_s - h_{vp}$$

$$= 6.36 + 0.3048 - 0.670$$

$$= 5.99 \text{ m}$$

804.0 *Related Calculations*
The following calculations can be used to determine conditions which affect pump performance.

804.1 Net Positive Suction Head Available, NPSH$_A$, should be compared with Net Positive Suction Head Required, NPSH$_R$, from the pump manufacturer's curve or data. Insufficient head can cause pumping problems.

804.2 *Effects of Speed and Impeller Diameter*
The effects of minor changes in speed or impeller diameter can be computed when test conditions vary from published data.

804.2.1 *Constant Speed*
Capacity varies in direct proportion to the impeller diameter:

$$Q_2 = Q_1 (D_2/D_1)$$

Total Head varies as the square of the ratio of the impeller diameters:

$$H_2 = H_1 (D_2/D_1)^2$$

Horsepower varies as the cube of the impeller diameter ratio:

$$Hp_2 = Hp_1 (D_2/D_1)^3$$

804.2.2 *Constant Impeller Diameter*
Capacity varies in direct proportion to the speed:

$$Q_2 = Q_1 (n_2/n_1)$$

Total Head varies as the square of the ratio of the speeds:

$$H_2 = H_1 (D_2/D_1)^2$$

Horsepower varies as the cube of the speed ratio:

$$Hp_2 = H_1 (D_2/D_1)^3$$

804.2.3 Where:

D = Impeller Diameter
H = Head
Q = Capacity
n = Rotational Speed
Hp = Power

805.0 References

805.1 Hydraulic Institute Standards for Centrifugal, Rotary and Reciprocating Pumps, 14th Edition. Available from Hydraulic Institute, 712 Lakewood Center North, 14600 Detroit Ave., Cleveland, OH 44107.

805.2 Cameron Hydraulic Data, 16th Edition. Available from Ingersoll-Rand Co., Woodcliff Lake, NJ 07675.

805.3 American National Standards/ASME Centrifugal Pumps Performance Test Code, PTC 8.2. Available from American Society of Mechanical Engineers, United Engineering Center, 345 East 47 St., New York, NY 10017.